Amateur Radio Station Log Book

Log Book

Start Date : ....

End Date : ....

Log Book No. : ....

Notes : ....

Personal Information

Name : ....

Phone : ....

Email : ....

Address : ....

Station

Call sign : ....

QTH Locator : ....

Equipment : ....

....

....

....

....

Antennas : ....

....

....

Amateur Radio Station Log

Station Call sign:

QRP - Mobile - Portable- Home

QTH:

Antenna:

Date	Time UTC		Frequency (MHz)	Station	Mode	Power	Report		QTH	QSL		
	Start	Finish					Sent	Rcvd		Sent	Rcvd	
												1
Notes												
												2
Notes												
												3
Notes												
												4
Notes												
												5
Notes												
												6
Notes												
												7
Notes												
												8
Notes												
												9
Notes												
												10
Notes												
												11
Notes												

Amateur Radio Station Log

Station Call sign: *QTH:*

QRP - Mobile - Portable- Home *Antenna:*

	Date	Time UTC		Frequency (MHz)	Station	Mode	Power	Report		QTH	QSL	
		Start	Finish					Sent	Rcvd		Sent	Rcvd
1												
	Notes											
2												
	Notes											
3												
	Notes											
4												
	Notes											
5												
	Notes											
6												
	Notes											
7												
	Notes											
8												
	Notes											
9												
	Notes											
10												
	Notes											
11												
	Notes											

Amateur Radio Station Log

Station Call sign:

QRP - Mobile - Portable- Home

QTH:

Antenna:

Date	Time UTC		Frequency (MHz)	Station	Mode	Power	Report		QTH	QSL		
	Start	Finish					Sent	Rcvd		Sent	Rcvd	
												1
Notes												
												2
Notes												
												3
Notes												
												4
Notes												
												5
Notes												
												6
Notes												
												7
Notes												
												8
Notes												
												9
Notes												
												10
Notes												
												11
Notes												

Amateur Radio Station Log

Station Call sign: *QTH:*

QRP - Mobile - Portable- Home *Antenna:*

	Date	Time UTC		Frequency (MHz)	Station	Mode	Power	Report		QTH	QSL	
		Start	Finish					Sent	Rcvd		Sent	Rcvd
1												
	Notes											
2												
	Notes											
3												
	Notes											
4												
	Notes											
5												
	Notes											
6												
	Notes											
7												
	Notes											
8												
	Notes											
9												
	Notes											
10												
	Notes											
11												
	Notes											

Amateur Radio Station Log

Station Call sign:

QRP - Mobile - Portable- Home

QTH:

Antenna:

Date	Time UTC		Frequency (MHz)	Station	Mode	Power	Report		QTH	QSL		
	Start	Finish					Sent	Rcvd		Sent	Rcvd	
												1
Notes												
												2
Notes												
												3
Notes												
												4
Notes												
												5
Notes												
												6
Notes												
												7
Notes												
												8
Notes												
												9
Notes												
												10
Notes												
												11
Notes												

Amateur Radio Station Log

Station Call sign:

QTH:

QRP - Mobile - Portable- Home

Antenna:

	Date	Time UTC		Frequency (MHz)	Station	Mode	Power	Report		QTH	QSL	
		Start	Finish					Sent	Rcvd		Sent	Rcvd
1												
	Notes											
2												
	Notes											
3												
	Notes											
4												
	Notes											
5												
	Notes											
6												
	Notes											
7												
	Notes											
8												
	Notes											
9												
	Notes											
10												
	Notes											
11												
	Notes											

Amateur Radio Station Log

Station Call sign:

QTH:

QRP - Mobile - Portable- Home

Antenna:

Date	Time UTC		Frequency (MHz)	Station	Mode	Power	Report		QTH	QSL		
	Start	Finish					Sent	Rcvd		Sent	Rcvd	
												1
Notes												
												2
Notes												
												3
Notes												
												4
Notes												
												5
Notes												
												6
Notes												
												7
Notes												
												8
Notes												
												9
Notes												
												10
Notes												
												11
Notes												

Amateur Radio Station Log

Station Call sign: *QTH:*

QRP - Mobile - Portable- Home *Antenna:*

	Date	Time UTC		Frequency (MHz)	Station	Mode	Power	Report		QTH	QSL	
		Start	Finish					Sent	Rcvd		Sent	Rcvd
1												
	Notes											
2												
	Notes											
3												
	Notes											
4												
	Notes											
5												
	Notes											
6												
	Notes											
7												
	Notes											
8												
	Notes											
9												
	Notes											
10												
	Notes											
11												
	Notes											

Amateur Radio Station Log

Station Call sign: *QTH:*

QRP - Mobile - Portable- Home *Antenna:*

Date	Time UTC		Frequency (MHz)	Station	Mode	Power	Report		QTH	QSL		
	Start	Finish					Sent	Rcvd		Sent	Rcvd	
												1
Notes												
												2
Notes												
												3
Notes												
												4
Notes												
												5
Notes												
												6
Notes												
												7
Notes												
												8
Notes												
												9
Notes												
												10
Notes												
												11
Notes												

Amateur Radio Station Log

Station Call sign:

QTH:

QRP - Mobile - Portable- Home

Antenna:

	Date	Time UTC		Frequency (MHz)	Station	Mode	Power	Report		QTH	QSL	
		Start	Finish					Sent	Rcvd		Sent	Rcvd
1												
	Notes											
2												
	Notes											
3												
	Notes											
4												
	Notes											
5												
	Notes											
6												
	Notes											
7												
	Notes											
8												
	Notes											
9												
	Notes											
10												
	Notes											
11												
	Notes											

Amateur Radio Station Log

Station Call sign:

QTH:

QRP - Mobile - Portable- Home

Antenna:

Date	Time UTC		Frequency (MHz)	Station	Mode	Power	Report		QTH	QSL		
	Start	Finish					Sent	Rcvd		Sent	Rcvd	
												1
Notes												
												2
Notes												
												3
Notes												
												4
Notes												
												5
Notes												
												6
Notes												
												7
Notes												
												8
Notes												
												9
Notes												
												10
Notes												
												11
Notes												

Amateur Radio Station Log

Station Call sign:

QRP - Mobile - Portable- Home

QTH:

Antenna:

	Date	Time UTC		Frequency (MHz)	Station	Mode	Power	Report		QTH	QSL	
		Start	Finish					Sent	Rcvd		Sent	Rcvd
1												
	Notes											
2												
	Notes											
3												
	Notes											
4												
	Notes											
5												
	Notes											
6												
	Notes											
7												
	Notes											
8												
	Notes											
9												
	Notes											
10												
	Notes											
11												
	Notes											

Amateur Radio Station Log

Station Call sign:

QRP - Mobile - Portable- Home

QTH:

Antenna:

Date	Time UTC		Frequency (MHz)	Station	Mode	Power	Report		QTH	QSL		
	Start	Finish					Sent	Rcvd		Sent	Rcvd	
												1
Notes												
												2
Notes												
												3
Notes												
												4
Notes												
												5
Notes												
												6
Notes												
												7
Notes												
												8
Notes												
												9
Notes												
												10
Notes												
												11
Notes												

Amateur Radio Station Log

Station Call sign:

QTH:

QRP - Mobile - Portable- Home

Antenna:

	Date	Time UTC		Frequency (MHz)	Station	Mode	Power	Report		QTH	QSL	
		Start	Finish					Sent	Rcvd		Sent	Rcvd
1												
	Notes											
2												
	Notes											
3												
	Notes											
4												
	Notes											
5												
	Notes											
6												
	Notes											
7												
	Notes											
8												
	Notes											
9												
	Notes											
10												
	Notes											
11												
	Notes											

Amateur Radio Station Log

Station Call sign:

QTH:

QRP - Mobile - Portable- Home

Antenna:

Date	Time UTC		Frequency (MHz)	Station	Mode	Power	Report		QTH	QSL		
	Start	Finish					Sent	Rcvd		Sent	Rcvd	
												1
Notes												
												2
Notes												
												3
Notes												
												4
Notes												
												5
Notes												
												6
Notes												
												7
Notes												
												8
Notes												
												9
Notes												
												10
Notes												
												11
Notes												

Amateur Radio Station Log

Station Call sign:

QTH:

QRP - Mobile - Portable- Home

Antenna:

	Date	Time UTC		Frequency (MHz)	Station	Mode	Power	Report		QTH	QSL	
		Start	Finish					Sent	Rcvd		Sent	Rcvd
1												
	Notes											
2												
	Notes											
3												
	Notes											
4												
	Notes											
5												
	Notes											
6												
	Notes											
7												
	Notes											
8												
	Notes											
9												
	Notes											
10												
	Notes											
11												
	Notes											

Amateur Radio Station Log

Station Call sign:

QRP - Mobile - Portable- Home

QTH:

Antenna:

Date	Time UTC		Frequency (MHz)	Station	Mode	Power	Report		QTH	QSL		
	Start	Finish					Sent	Rcvd		Sent	Rcvd	
												1
Notes												
												2
Notes												
												3
Notes												
												4
Notes												
												5
Notes												
												6
Notes												
												7
Notes												
												8
Notes												
												9
Notes												
												10
Notes												
												11
Notes												

Amateur Radio Station Log

Station Call sign:

QTH:

QRP - Mobile - Portable- Home

Antenna:

	Date	Time UTC		Frequency (MHz)	Station	Mode	Power	Report		QTH	QSL	
		Start	Finish					Sent	Rcvd		Sent	Rcvd
1												
	Notes											
2												
	Notes											
3												
	Notes											
4												
	Notes											
5												
	Notes											
6												
	Notes											
7												
	Notes											
8												
	Notes											
9												
	Notes											
10												
	Notes											
11												
	Notes											

Amateur Radio Station Log

Station Call sign:

QTH:

QRP - Mobile - Portable- Home

Antenna:

Date	Time UTC		Frequency (MHz)	Station	Mode	Power	Report		QTH	QSL		
	Start	Finish					Sent	Rcvd		Sent	Rcvd	
												1
Notes												
												2
Notes												
												3
Notes												
												4
Notes												
												5
Notes												
												6
Notes												
												7
Notes												
												8
Notes												
												9
Notes												
												10
Notes												
												11
Notes												

Amateur Radio Station Log

Station Call sign:

QTH:

QRP - Mobile - Portable- Home

Antenna:

	Date	Time UTC		Frequency (MHz)	Station	Mode	Power	Report		QTH	QSL	
		Start	Finish					Sent	Rcvd		Sent	Rcvd
1												
	Notes											
2												
	Notes											
3												
	Notes											
4												
	Notes											
5												
	Notes											
6												
	Notes											
7												
	Notes											
8												
	Notes											
9												
	Notes											
10												
	Notes											
11												
	Notes											

Amateur Radio Station Log

Station Call sign:

QTH:

QRP - Mobile - Portable- Home

Antenna:

Date	Time UTC		Frequency (MHz)	Station	Mode	Power	Report		QTH	QSL		
	Start	Finish					Sent	Rcvd		Sent	Rcvd	
												1
Notes												
												2
Notes												
												3
Notes												
												4
Notes												
												5
Notes												
												6
Notes												
												7
Notes												
												8
Notes												
												9
Notes												
												10
Notes												
												11
Notes												

Amateur Radio Station Log

Station Call sign:

QTH:

QRP - Mobile - Portable- Home

Antenna:

	Date	Time UTC		Frequency (MHz)	Station	Mode	Power	Report		QTH	QSL	
		Start	Finish					Sent	Rcvd		Sent	Rcvd
1												
	Notes											
2												
	Notes											
3												
	Notes											
4												
	Notes											
5												
	Notes											
6												
	Notes											
7												
	Notes											
8												
	Notes											
9												
	Notes											
10												
	Notes											
11												
	Notes											

Amateur Radio Station Log

Station Call sign:

QTH:

QRP - Mobile - Portable- Home

Antenna:

Date	Time UTC		Frequency (MHz)	Station	Mode	Power	Report		QTH	QSL		
	Start	Finish					Sent	Rcvd		Sent	Rcvd	
												1
Notes												
												2
Notes												
												3
Notes												
												4
Notes												
												5
Notes												
												6
Notes												
												7
Notes												
												8
Notes												
												9
Notes												
												10
Notes												
												11
Notes												

Amateur Radio Station Log

Station Call sign:

QTH:

QRP - Mobile - Portable- Home

Antenna:

	Date	Time UTC		Frequency (MHz)	Station	Mode	Power	Report		QTH	QSL	
		Start	Finish					Sent	Rcvd		Sent	Rcvd
1												
	Notes											
2												
	Notes											
3												
	Notes											
4												
	Notes											
5												
	Notes											
6												
	Notes											
7												
	Notes											
8												
	Notes											
9												
	Notes											
10												
	Notes											
11												
	Notes											

Amateur Radio Station Log

Station Call sign:

QRP - Mobile - Portable- Home

QTH:

Antenna:

Date	Time UTC		Frequency (MHz)	Station	Mode	Power	Report		QTH	QSL		
	Start	Finish					Sent	Rcvd		Sent	Rcvd	
												1
Notes												
												2
Notes												
												3
Notes												
												4
Notes												
												5
Notes												
												6
Notes												
												7
Notes												
												8
Notes												
												9
Notes												
												10
Notes												
												11
Notes												

Amateur Radio Station Log

Station Call sign:

QRP - Mobile - Portable- Home

QTH:

Antenna:

	Date	Time UTC		Frequency (MHz)	Station	Mode	Power	Report		QTH	QSL	
		Start	Finish					Sent	Rcvd		Sent	Rcvd
1												
	Notes											
2												
	Notes											
3												
	Notes											
4												
	Notes											
5												
	Notes											
6												
	Notes											
7												
	Notes											
8												
	Notes											
9												
	Notes											
10												
	Notes											
11												
	Notes											

Amateur Radio Station Log

Station Call sign:

QTH:

QRP - Mobile - Portable- Home

Antenna:

Date	Time UTC		Frequency (MHz)	Station	Mode	Power	Report		QTH	QSL		
	Start	Finish					Sent	Rcvd		Sent	Rcvd	
												1
Notes												
												2
Notes												
												3
Notes												
												4
Notes												
												5
Notes												
												6
Notes												
												7
Notes												
												8
Notes												
												9
Notes												
												10
Notes												
												11
Notes												

Amateur Radio Station Log

Station Call sign: | ***QTH:***

QRP - Mobile - Portable- Home | ***Antenna:***

	Date	Time UTC		Frequency (MHz)	Station	Mode	Power	Report		QTH	QSL	
		Start	Finish					Sent	Rcvd		Sent	Rcvd
1												
	Notes											
2												
	Notes											
3												
	Notes											
4												
	Notes											
5												
	Notes											
6												
	Notes											
7												
	Notes											
8												
	Notes											
9												
	Notes											
10												
	Notes											
11												
	Notes											

Amateur Radio Station Log

Station Call sign:

QRP - Mobile - Portable- Home

QTH:

Antenna:

Date	Time UTC		Frequency (MHz)	Station	Mode	Power	Report		QTH	QSL		
	Start	Finish					Sent	Rcvd		Sent	Rcvd	
												1
Notes												
												2
Notes												
												3
Notes												
												4
Notes												
												5
Notes												
												6
Notes												
												7
Notes												
												8
Notes												
												9
Notes												
												10
Notes												
												11
Notes												

Amateur Radio Station Log

Station Call sign:

QTH:

QRP - Mobile - Portable- Home

Antenna:

Date	Time UTC		Frequency (MHz)	Station	Mode	Power	Report		QTH	QSL	
	Start	Finish					Sent	Rcvd		Sent	Rcvd
1											
Notes											
2											
Notes											
3											
Notes											
4											
Notes											
5											
Notes											
6											
Notes											
7											
Notes											
8											
Notes											
9											
Notes											
10											
Notes											
11											
Notes											

Amateur Radio Station Log

Station Call sign:

QTH:

QRP - Mobile - Portable- Home

Antenna:

Date	Time UTC		Frequency (MHz)	Station	Mode	Power	Report		QTH	QSL		
	Start	Finish					Sent	Rcvd		Sent	Rcvd	
												1
Notes												
												2
Notes												
												3
Notes												
												4
Notes												
												5
Notes												
												6
Notes												
												7
Notes												
												8
Notes												
												9
Notes												
												10
Notes												
												11
Notes												

Amateur Radio Station Log

Station Call sign:

QTH:

QRP - Mobile - Portable- Home

Antenna:

	Date	Time UTC		Frequency (MHz)	Station	Mode	Power	Report		QTH	QSL	
		Start	Finish					Sent	Rcvd		Sent	Rcvd
1												
	Notes											
2												
	Notes											
3												
	Notes											
4												
	Notes											
5												
	Notes											
6												
	Notes											
7												
	Notes											
8												
	Notes											
9												
	Notes											
10												
	Notes											
11												
	Notes											

Amateur Radio Station Log

Station Call sign:

QTH:

QRP - Mobile - Portable- Home

Antenna:

Date	Time UTC		Frequency (MHz)	Station	Mode	Power	Report		QTH	QSL		
	Start	Finish					Sent	Rcvd		Sent	Rcvd	
												1
Notes												
												2
Notes												
												3
Notes												
												4
Notes												
												5
Notes												
												6
Notes												
												7
Notes												
												8
Notes												
												9
Notes												
												10
Notes												
												11
Notes												

Amateur Radio Station Log

Station Call sign:

QTH:

QRP - Mobile - Portable- Home

Antenna:

	Date	Time UTC		Frequency (MHz)	Station	Mode	Power	Report		QTH	QSL	
		Start	Finish					Sent	Rcvd		Sent	Rcvd
1												
	Notes											
2												
	Notes											
3												
	Notes											
4												
	Notes											
5												
	Notes											
6												
	Notes											
7												
	Notes											
8												
	Notes											
9												
	Notes											
10												
	Notes											
11												
	Notes											

Amateur Radio Station Log

Station Call sign:

QRP - Mobile - Portable- Home

QTH:

Antenna:

Date	Time UTC		Frequency (MHz)	Station	Mode	Power	Report		QTH	QSL		
	Start	Finish					Sent	Rcvd		Sent	Rcvd	
												1
Notes												
												2
Notes												
												3
Notes												
												4
Notes												
												5
Notes												
												6
Notes												
												7
Notes												
												8
Notes												
												9
Notes												
												10
Notes												
												11
Notes												

Amateur Radio Station Log

Station Call sign: *QTH:*

QRP - Mobile - Portable- Home *Antenna:*

	Date	Time UTC		Frequency (MHz)	Station	Mode	Power	Report		QTH	QSL	
		Start	Finish					Sent	Rcvd		Sent	Rcvd
1												
	Notes											
2												
	Notes											
3												
	Notes											
4												
	Notes											
5												
	Notes											
6												
	Notes											
7												
	Notes											
8												
	Notes											
9												
	Notes											
10												
	Notes											
11												
	Notes											

Amateur Radio Station Log

Station Call sign:

QRP - Mobile - Portable- Home

QTH:

Antenna:

Date	Time UTC		Frequency (MHz)	Station	Mode	Power	Report		QTH	QSL		
	Start	Finish					Sent	Rcvd		Sent	Rcvd	
												1
Notes												
												2
Notes												
												3
Notes												
												4
Notes												
												5
Notes												
												6
Notes												
												7
Notes												
												8
Notes												
												9
Notes												
												10
Notes												
												11
Notes												

Amateur Radio Station Log

Station Call sign:

QTH:

QRP - Mobile - Portable- Home

Antenna:

	Date	Time UTC		Frequency (MHz)	Station	Mode	Power	Report		QTH	QSL	
		Start	Finish					Sent	Rcvd		Sent	Rcvd
1												
	Notes											
2												
	Notes											
3												
	Notes											
4												
	Notes											
5												
	Notes											
6												
	Notes											
7												
	Notes											
8												
	Notes											
9												
	Notes											
10												
	Notes											
11												
	Notes											

Amateur Radio Station Log

Station Call sign:

QRP - Mobile - Portable- Home

QTH:

Antenna:

Date	Time UTC		Frequency (MHz)	Station	Mode	Power	Report		QTH	QSL		
	Start	Finish					Sent	Rcvd		Sent	Rcvd	
												1
Notes												
												2
Notes												
												3
Notes												
												4
Notes												
												5
Notes												
												6
Notes												
												7
Notes												
												8
Notes												
												9
Notes												
												10
Notes												
												11
Notes												

Amateur Radio Station Log

Station Call sign:

QTH:

QRP - Mobile - Portable- Home

Antenna:

Date	Time UTC		Frequency (MHz)	Station	Mode	Power	Report		QTH	QSL	
	Start	Finish					Sent	Rcvd		Sent	Rcvd
1											
Notes											
2											
Notes											
3											
Notes											
4											
Notes											
5											
Notes											
6											
Notes											
7											
Notes											
8											
Notes											
9											
Notes											
10											
Notes											
11											
Notes											

Amateur Radio Station Log

Station Call sign:

QRP - Mobile - Portable- Home

QTH:

Antenna:

Date	Time UTC		Frequency (MHz)	Station	Mode	Power	Report		QTH	QSL		
	Start	Finish					Sent	Rcvd		Sent	Rcvd	
												1
Notes												
												2
Notes												
												3
Notes												
												4
Notes												
												5
Notes												
												6
Notes												
												7
Notes												
												8
Notes												
												9
Notes												
												10
Notes												
												11
Notes												

Amateur Radio Station Log

Station Call sign:

QTH:

QRP - Mobile - Portable- Home

Antenna:

	Date	Time UTC		Frequency (MHz)	Station	Mode	Power	Report		QTH	QSL	
		Start	Finish					Sent	Rcvd		Sent	Rcvd
1												
	Notes											
2												
	Notes											
3												
	Notes											
4												
	Notes											
5												
	Notes											
6												
	Notes											
7												
	Notes											
8												
	Notes											
9												
	Notes											
10												
	Notes											
11												
	Notes											

Amateur Radio Station Log

Station Call sign:

QTH:

QRP - Mobile - Portable- Home

Antenna:

Date	Time UTC		Frequency (MHz)	Station	Mode	Power	Report		QTH	QSL		
	Start	Finish					Sent	Rcvd		Sent	Rcvd	
												1
Notes												
												2
Notes												
												3
Notes												
												4
Notes												
												5
Notes												
												6
Notes												
												7
Notes												
												8
Notes												
												9
Notes												
												10
Notes												
												11
Notes												

Amateur Radio Station Log

Station Call sign:

QTH:

QRP - Mobile - Portable- Home

Antenna:

Date	Time UTC		Frequency (MHz)	Station	Mode	Power	Report		QTH	QSL	
	Start	Finish					Sent	Rcvd		Sent	Rcvd
1											
Notes											
2											
Notes											
3											
Notes											
4											
Notes											
5											
Notes											
6											
Notes											
7											
Notes											
8											
Notes											
9											
Notes											
10											
Notes											
11											
Notes											

Amateur Radio Station Log

Station Call sign:

QRP - Mobile - Portable- Home

QTH:

Antenna:

Date	Time UTC		Frequency (MHz)	Station	Mode	Power	Report		QTH	QSL		
	Start	Finish					Sent	Rcvd		Sent	Rcvd	
												1
Notes												
												2
Notes												
												3
Notes												
												4
Notes												
												5
Notes												
												6
Notes												
												7
Notes												
												8
Notes												
												9
Notes												
												10
Notes												
												11
Notes												

Amateur Radio Station Log

Station Call sign:

QTH:

QRP - Mobile - Portable- Home

Antenna:

	Date	Time UTC		Frequency (MHz)	Station	Mode	Power	Report		QTH	QSL	
		Start	Finish					Sent	Rcvd		Sent	Rcvd
1												
	Notes											
2												
	Notes											
3												
	Notes											
4												
	Notes											
5												
	Notes											
6												
	Notes											
7												
	Notes											
8												
	Notes											
9												
	Notes											
10												
	Notes											
11												
	Notes											

Amateur Radio Station Log

Station Call sign:

QRP - Mobile - Portable- Home

QTH:

Antenna:

Date	Time UTC		Frequency (MHz)	Station	Mode	Power	Report		QTH	QSL		
	Start	Finish					Sent	Rcvd		Sent	Rcvd	
												1
Notes												
												2
Notes												
												3
Notes												
												4
Notes												
												5
Notes												
												6
Notes												
												7
Notes												
												8
Notes												
												9
Notes												
												10
Notes												
												11
Notes												

Amateur Radio Station Log

Station Call sign:

QTH:

QRP - Mobile - Portable- Home

Antenna:

	Date	Time UTC		Frequency (MHz)	Station	Mode	Power	Report		QTH	QSL	
		Start	Finish					Sent	Rcvd		Sent	Rcvd
1												
	Notes											
2												
	Notes											
3												
	Notes											
4												
	Notes											
5												
	Notes											
6												
	Notes											
7												
	Notes											
8												
	Notes											
9												
	Notes											
10												
	Notes											
11												
	Notes											

Amateur Radio Station Log

Station Call sign:

QTH:

QRP - Mobile - Portable- Home

Antenna:

Date	Time UTC		Frequency (MHz)	Station	Mode	Power	Report		QTH	QSL		
	Start	Finish					Sent	Rcvd		Sent	Rcvd	
												1
Notes												
												2
Notes												
												3
Notes												
												4
Notes												
												5
Notes												
												6
Notes												
												7
Notes												
												8
Notes												
												9
Notes												
												10
Notes												
												11
Notes												

Amateur Radio Station Log

Station Call sign:

QTH:

QRP - Mobile - Portable- Home

Antenna:

Date	Time UTC		Frequency (MHz)	Station	Mode	Power	Report		QTH	QSL	
	Start	Finish					Sent	Rcvd		Sent	Rcvd
1											
Notes											
2											
Notes											
3											
Notes											
4											
Notes											
5											
Notes											
6											
Notes											
7											
Notes											
8											
Notes											
9											
Notes											
10											
Notes											
11											
Notes											

Amateur Radio Station Log

Station Call sign:

QTH:

QRP - Mobile - Portable- Home

Antenna:

Date	Time UTC		Frequency (MHz)	Station	Mode	Power	Report		QTH	QSL		
	Start	Finish					Sent	Rcvd		Sent	Rcvd	
												1
Notes												
												2
Notes												
												3
Notes												
												4
Notes												
												5
Notes												
												6
Notes												
												7
Notes												
												8
Notes												
												9
Notes												
												10
Notes												
												11
Notes												

Amateur Radio Station Log

Station Call sign:

QRP - Mobile - Portable- Home

QTH:

Antenna:

	Date	Time UTC		Frequency (MHz)	Station	Mode	Power	Report		QTH	QSL	
		Start	Finish					Sent	Rcvd		Sent	Rcvd
1												
	Notes											
2												
	Notes											
3												
	Notes											
4												
	Notes											
5												
	Notes											
6												
	Notes											
7												
	Notes											
8												
	Notes											
9												
	Notes											
10												
	Notes											
11												
	Notes											

Amateur Radio Station Log

Station Call sign:

QTH:

QRP - Mobile - Portable- Home

Antenna:

Date	Time UTC		Frequency (MHz)	Station	Mode	Power	Report		QTH	QSL		
	Start	Finish					Sent	Rcvd		Sent	Rcvd	
												1
Notes												
												2
Notes												
												3
Notes												
												4
Notes												
												5
Notes												
												6
Notes												
												7
Notes												
												8
Notes												
												9
Notes												
												10
Notes												
												11
Notes												

Amateur Radio Station Log

Station Call sign:

QRP - Mobile - Portable- Home

QTH:

Antenna:

	Date	Time UTC		Frequency (MHz)	Station	Mode	Power	Report		QTH	QSL	
		Start	Finish					Sent	Rcvd		Sent	Rcvd
1												
	Notes											
2												
	Notes											
3												
	Notes											
4												
	Notes											
5												
	Notes											
6												
	Notes											
7												
	Notes											
8												
	Notes											
9												
	Notes											
10												
	Notes											
11												
	Notes											

Amateur Radio Station Log

Station Call sign:

QRP - Mobile - Portable- Home

QTH:

Antenna:

Date	Time UTC		Frequency (MHz)	Station	Mode	Power	Report		QTH	QSL		
	Start	Finish					Sent	Rcvd		Sent	Rcvd	
												1
Notes												
												2
Notes												
												3
Notes												
												4
Notes												
												5
Notes												
												6
Notes												
												7
Notes												
												8
Notes												
												9
Notes												
												10
Notes												
												11
Notes												

Amateur Radio Station Log

Station Call sign:

QTH:

QRP - Mobile - Portable- Home

Antenna:

	Date	Time UTC		Frequency (MHz)	Station	Mode	Power	Report		QTH	QSL	
		Start	Finish					Sent	Rcvd		Sent	Rcvd
1												
	Notes											
2												
	Notes											
3												
	Notes											
4												
	Notes											
5												
	Notes											
6												
	Notes											
7												
	Notes											
8												
	Notes											
9												
	Notes											
10												
	Notes											
11												
	Notes											

Amateur Radio Station Log

Station Call sign:

QRP - Mobile - Portable- Home

QTH:

Antenna:

Date	Time UTC		Frequency (MHz)	Station	Mode	Power	Report		QTH	QSL		
	Start	Finish					Sent	Rcvd		Sent	Rcvd	
												1
Notes												
												2
Notes												
												3
Notes												
												4
Notes												
												5
Notes												
												6
Notes												
												7
Notes												
												8
Notes												
												9
Notes												
												10
Notes												
												11
Notes												

Amateur Radio Station Log

Station Call sign:

QTH:

QRP - Mobile - Portable- Home

Antenna:

	Date	Time UTC		Frequency (MHz)	Station	Mode	Power	Report		QTH	QSL	
		Start	Finish					Sent	Rcvd		Sent	Rcvd
1												
	Notes											
2												
	Notes											
3												
	Notes											
4												
	Notes											
5												
	Notes											
6												
	Notes											
7												
	Notes											
8												
	Notes											
9												
	Notes											
10												
	Notes											
11												
	Notes											

Amateur Radio Station Log

Station Call sign:

QRP - Mobile - Portable- Home

QTH:

Antenna:

Date	Time UTC		Frequency (MHz)	Station	Mode	Power	Report		QTH	QSL		
	Start	Finish					Sent	Rcvd		Sent	Rcvd	
												1
Notes												
												2
Notes												
												3
Notes												
												4
Notes												
												5
Notes												
												6
Notes												
												7
Notes												
												8
Notes												
												9
Notes												
												10
Notes												
												11
Notes												

Amateur Radio Station Log

Station Call sign:

QRP - Mobile - Portable- Home

QTH:

Antenna:

	Date	Time UTC		Frequency (MHz)	Station	Mode	Power	Report		QTH	QSL	
		Start	Finish					Sent	Rcvd		Sent	Rcvd
1												
	Notes											
2												
	Notes											
3												
	Notes											
4												
	Notes											
5												
	Notes											
6												
	Notes											
7												
	Notes											
8												
	Notes											
9												
	Notes											
10												
	Notes											
11												
	Notes											

Amateur Radio Station Log

Station Call sign:

QTH:

QRP - Mobile - Portable- Home

Antenna:

Date	Time UTC		Frequency (MHz)	Station	Mode	Power	Report		QTH	QSL		
	Start	Finish					Sent	Rcvd		Sent	Rcvd	
												1
Notes												
												2
Notes												
												3
Notes												
												4
Notes												
												5
Notes												
												6
Notes												
												7
Notes												
												8
Notes												
												9
Notes												
												10
Notes												
												11
Notes												

Amateur Radio Station Log

Station Call sign:

QTH:

QRP - Mobile - Portable- Home

Antenna:

	Date	Time UTC		Frequency (MHz)	Station	Mode	Power	Report		QTH	QSL	
		Start	Finish					Sent	Rcvd		Sent	Rcvd
1												
	Notes											
2												
	Notes											
3												
	Notes											
4												
	Notes											
5												
	Notes											
6												
	Notes											
7												
	Notes											
8												
	Notes											
9												
	Notes											
10												
	Notes											
11												
	Notes											

Amateur Radio Station Log

Station Call sign:

QTH:

QRP - Mobile - Portable- Home

Antenna:

Date	Time UTC		Frequency (MHz)	Station	Mode	Power	Report		QTH	QSL		
	Start	Finish					Sent	Rcvd		Sent	Rcvd	
												1
Notes												
												2
Notes												
												3
Notes												
												4
Notes												
												5
Notes												
												6
Notes												
												7
Notes												
												8
Notes												
												9
Notes												
												10
Notes												
												11
Notes												

Amateur Radio Station Log

Station Call sign: QTH:

QRP - Mobile - Portable- Home Antenna:

	Date	Time UTC		Frequency (MHz)	Station	Mode	Power	Report		QTH	QSL	
		Start	Finish					Sent	Rcvd		Sent	Rcvd
1												
	Notes											
2												
	Notes											
3												
	Notes											
4												
	Notes											
5												
	Notes											
6												
	Notes											
7												
	Notes											
8												
	Notes											
9												
	Notes											
10												
	Notes											
11												
	Notes											

Amateur Radio Station Log

Station Call sign:

QTH:

QRP - Mobile - Portable- Home

Antenna:

Date	Time UTC		Frequency (MHz)	Station	Mode	Power	Report		QTH	QSL		
	Start	Finish					Sent	Rcvd		Sent	Rcvd	
												1
Notes												
												2
Notes												
												3
Notes												
												4
Notes												
												5
Notes												
												6
Notes												
												7
Notes												
												8
Notes												
												9
Notes												
												10
Notes												
												11
Notes												

Amateur Radio Station Log

Station Call sign:

QTH:

QRP - Mobile - Portable- Home

Antenna:

	Date	Time UTC		Frequency (MHz)	Station	Mode	Power	Report		QTH	QSL	
		Start	Finish					Sent	Rcvd		Sent	Rcvd
1												
	Notes											
2												
	Notes											
3												
	Notes											
4												
	Notes											
5												
	Notes											
6												
	Notes											
7												
	Notes											
8												
	Notes											
9												
	Notes											
10												
	Notes											
11												
	Notes											

Amateur Radio Station Log

Station Call sign:

QRP - Mobile - Portable- Home

QTH:

Antenna:

Date	Time UTC		Frequency (MHz)	Station	Mode	Power	Report		QTH	QSL		
	Start	Finish					Sent	Rcvd		Sent	Rcvd	
												1
Notes												
												2
Notes												
												3
Notes												
												4
Notes												
												5
Notes												
												6
Notes												
												7
Notes												
												8
Notes												
												9
Notes												
												10
Notes												
												11
Notes												

Amateur Radio Station Log

Station Call sign:

QTH:

QRP - Mobile - Portable- Home

Antenna:

	Date	Time UTC		Frequency (MHz)	Station	Mode	Power	Report		QTH	QSL	
		Start	Finish					Sent	Rcvd		Sent	Rcvd
1												
	Notes											
2												
	Notes											
3												
	Notes											
4												
	Notes											
5												
	Notes											
6												
	Notes											
7												
	Notes											
8												
	Notes											
9												
	Notes											
10												
	Notes											
11												
	Notes											

Amateur Radio Station Log

Station Call sign:

QRP - Mobile - Portable- Home

QTH:

Antenna:

Date	Time UTC		Frequency (MHz)	Station	Mode	Power	Report		QTH	QSL		
	Start	Finish					Sent	Rcvd		Sent	Rcvd	
												1
Notes												
												2
Notes												
												3
Notes												
												4
Notes												
												5
Notes												
												6
Notes												
												7
Notes												
												8
Notes												
												9
Notes												
												10
Notes												
												11
Notes												

Amateur Radio Station Log

Station Call sign:

QTH:

QRP - Mobile - Portable- Home

Antenna:

Date	Time UTC		Frequency (MHz)	Station	Mode	Power	Report		QTH	QSL	
	Start	Finish					Sent	Rcvd		Sent	Rcvd
1											
Notes											
2											
Notes											
3											
Notes											
4											
Notes											
5											
Notes											
6											
Notes											
7											
Notes											
8											
Notes											
9											
Notes											
10											
Notes											
11											
Notes											

Amateur Radio Station Log

Station Call sign:

QRP - Mobile - Portable- Home

QTH:

Antenna:

Date	Time UTC		Frequency (MHz)	Station	Mode	Power	Report		QTH	QSL		
	Start	Finish					Sent	Rcvd		Sent	Rcvd	
												1
Notes												
												2
Notes												
												3
Notes												
												4
Notes												
												5
Notes												
												6
Notes												
												7
Notes												
												8
Notes												
												9
Notes												
												10
Notes												
												11
Notes												

Amateur Radio Station Log

Station Call sign:

QRP - Mobile - Portable- Home

QTH:

Antenna:

Date	Time UTC		Frequency (MHz)	Station	Mode	Power	Report		QTH	QSL	
	Start	Finish					Sent	Rcvd		Sent	Rcvd
1											
Notes											
2											
Notes											
3											
Notes											
4											
Notes											
5											
Notes											
6											
Notes											
7											
Notes											
8											
Notes											
9											
Notes											
10											
Notes											
11											
Notes											

Amateur Radio Station Log

Station Call sign:

QRP - Mobile - Portable- Home

QTH:

Antenna:

Date	Time UTC		Frequency (MHz)	Station	Mode	Power	Report		QTH	QSL		
	Start	Finish					Sent	Rcvd		Sent	Rcvd	
												1
Notes												
												2
Notes												
												3
Notes												
												4
Notes												
												5
Notes												
												6
Notes												
												7
Notes												
												8
Notes												
												9
Notes												
												10
Notes												
												11
Notes												

Amateur Radio Station Log

Station Call sign:

QTH:

QRP - Mobile - Portable- Home

Antenna:

	Date	Time UTC		Frequency (MHz)	Station	Mode	Power	Report		QTH	QSL	
		Start	Finish					Sent	Rcvd		Sent	Rcvd
1												
	Notes											
2												
	Notes											
3												
	Notes											
4												
	Notes											
5												
	Notes											
6												
	Notes											
7												
	Notes											
8												
	Notes											
9												
	Notes											
10												
	Notes											
11												
	Notes											

Amateur Radio Station Log

Station Call sign:

QTH:

QRP - Mobile - Portable- Home

Antenna:

Date	Time UTC		Frequency (MHz)	Station	Mode	Power	Report		QTH	QSL		
	Start	Finish					Sent	Rcvd		Sent	Rcvd	
												1
Notes												
												2
Notes												
												3
Notes												
												4
Notes												
												5
Notes												
												6
Notes												
												7
Notes												
												8
Notes												
												9
Notes												
												10
Notes												
												11
Notes												

Amateur Radio Station Log

Station Call sign:

QTH:

QRP - Mobile - Portable- Home

Antenna:

	Date	Time UTC		Frequency (MHz)	Station	Mode	Power	Report		QTH	QSL	
		Start	Finish					Sent	Rcvd		Sent	Rcvd
1												
	Notes											
2												
	Notes											
3												
	Notes											
4												
	Notes											
5												
	Notes											
6												
	Notes											
7												
	Notes											
8												
	Notes											
9												
	Notes											
10												
	Notes											
11												
	Notes											

Amateur Radio Station Log

Station Call sign:

QTH:

QRP - Mobile - Portable- Home

Antenna:

Date	Time UTC		Frequency (MHz)	Station	Mode	Power	Report		QTH	QSL		
	Start	Finish					Sent	Rcvd		Sent	Rcvd	
												1
Notes												
												2
Notes												
												3
Notes												
												4
Notes												
												5
Notes												
												6
Notes												
												7
Notes												
												8
Notes												
												9
Notes												
												10
Notes												
												11
Notes												

Amateur Radio Station Log

Station Call sign:

QTH:

QRP - Mobile - Portable- Home

Antenna:

	Date	Time UTC		Frequency (MHz)	Station	Mode	Power	Report		QTH	QSL	
		Start	Finish					Sent	Rcvd		Sent	Rcvd
1												
	Notes											
2												
	Notes											
3												
	Notes											
4												
	Notes											
5												
	Notes											
6												
	Notes											
7												
	Notes											
8												
	Notes											
9												
	Notes											
10												
	Notes											
11												
	Notes											

Amateur Radio Station Log

Station Call sign: *QTH:*

QRP - Mobile - Portable- Home *Antenna:*

Date	Time UTC		Frequency (MHz)	Station	Mode	Power	Report		QTH	QSL		
	Start	Finish					Sent	Rcvd		Sent	Rcvd	
												1
Notes												
												2
Notes												
												3
Notes												
												4
Notes												
												5
Notes												
												6
Notes												
												7
Notes												
												8
Notes												
												9
Notes												
												10
Notes												
												11
Notes												

Amateur Radio Station Log

Station Call sign:

QTH:

QRP - Mobile - Portable- Home

Antenna:

	Date	Time UTC		Frequency (MHz)	Station	Mode	Power	Report		QTH	QSL	
		Start	Finish					Sent	Rcvd		Sent	Rcvd
1												
	Notes											
2												
	Notes											
3												
	Notes											
4												
	Notes											
5												
	Notes											
6												
	Notes											
7												
	Notes											
8												
	Notes											
9												
	Notes											
10												
	Notes											
11												
	Notes											

Amateur Radio Station Log

Station Call sign:

QRP - Mobile - Portable- Home

QTH:

Antenna:

Date	Time UTC		Frequency (MHz)	Station	Mode	Power	Report		QTH	QSL	
	Start	Finish					Sent	Rcvd		Sent	Rcvd
												1
Notes											
												2
Notes											
												3
Notes											
												4
Notes											
												5
Notes											
												6
Notes											
												7
Notes											
												8
Notes											
												9
Notes											
												10
Notes											
												11
Notes											

Amateur Radio Station Log

Station Call sign:

QTH:

QRP - Mobile - Portable- Home

Antenna:

Date	Time UTC		Frequency (MHz)	Station	Mode	Power	Report		QTH	QSL	
	Start	Finish					Sent	Rcvd		Sent	Rcvd
1											
Notes											
2											
Notes											
3											
Notes											
4											
Notes											
5											
Notes											
6											
Notes											
7											
Notes											
8											
Notes											
9											
Notes											
10											
Notes											
11											
Notes											

Amateur Radio Station Log

Station Call sign:

QRP - Mobile - Portable- Home

QTH:

Antenna:

Date	Time UTC		Frequency (MHz)	Station	Mode	Power	Report		QTH	QSL		
	Start	Finish					Sent	Rcvd		Sent	Rcvd	
												1
Notes												
												2
Notes												
												3
Notes												
												4
Notes												
												5
Notes												
												6
Notes												
												7
Notes												
												8
Notes												
												9
Notes												
												10
Notes												
												11
Notes												

Amateur Radio Station Log

Station Call sign:

QRP - Mobile - Portable- Home

QTH:

Antenna:

	Date	Time UTC		Frequency (MHz)	Station	Mode	Power	Report		QTH	QSL	
		Start	Finish					Sent	Rcvd		Sent	Rcvd
1												
	Notes											
2												
	Notes											
3												
	Notes											
4												
	Notes											
5												
	Notes											
6												
	Notes											
7												
	Notes											
8												
	Notes											
9												
	Notes											
10												
	Notes											
11												
	Notes											

Amateur Radio Station Log

Station Call sign: *QTH:*

QRP - Mobile - Portable- Home *Antenna:*

Date	Time UTC		Frequency (MHz)	Station	Mode	Power	Report		QTH	QSL		
	Start	Finish					Sent	Rcvd		Sent	Rcvd	
												1
Notes												
												2
Notes												
												3
Notes												
												4
Notes												
												5
Notes												
												6
Notes												
												7
Notes												
												8
Notes												
												9
Notes												
												10
Notes												
												11
Notes												

Amateur Radio Station Log

Station Call sign:

QTH:

QRP - Mobile - Portable- Home

Antenna:

	Date	Time UTC		Frequency (MHz)	Station	Mode	Power	Report		QTH	QSL	
		Start	Finish					Sent	Rcvd		Sent	Rcvd
1												
	Notes											
2												
	Notes											
3												
	Notes											
4												
	Notes											
5												
	Notes											
6												
	Notes											
7												
	Notes											
8												
	Notes											
9												
	Notes											
10												
	Notes											
11												
	Notes											

Amateur Radio Station Log

Station Call sign:

QTH:

QRP - Mobile - Portable- Home

Antenna:

Date	Time UTC		Frequency (MHz)	Station	Mode	Power	Report		QTH	QSL		
	Start	Finish					Sent	Rcvd		Sent	Rcvd	
												1
Notes												
												2
Notes												
												3
Notes												
												4
Notes												
												5
Notes												
												6
Notes												
												7
Notes												
												8
Notes												
												9
Notes												
												10
Notes												
												11
Notes												

Amateur Radio Station Log

Station Call sign:

QTH:

QRP - Mobile - Portable- Home

Antenna:

	Date	Time UTC		Frequency (MHz)	Station	Mode	Power	Report		QTH	QSL	
		Start	Finish					Sent	Rcvd		Sent	Rcvd
1												
	Notes											
2												
	Notes											
3												
	Notes											
4												
	Notes											
5												
	Notes											
6												
	Notes											
7												
	Notes											
8												
	Notes											
9												
	Notes											
10												
	Notes											
11												
	Notes											

Amateur Radio Station Log

Station Call sign:

QRP - Mobile - Portable- Home

QTH:

Antenna:

Date	Time UTC		Frequency (MHz)	Station	Mode	Power	Report		QTH	QSL		
	Start	Finish					Sent	Rcvd		Sent	Rcvd	
												1
Notes												
												2
Notes												
												3
Notes												
												4
Notes												
												5
Notes												
												6
Notes												
												7
Notes												
												8
Notes												
												9
Notes												
												10
Notes												
												11
Notes												

Amateur Radio Station Log

Station Call sign:

QRP - Mobile - Portable- Home

QTH:

Antenna:

	Date	Time UTC		Frequency (MHz)	Station	Mode	Power	Report		QTH	QSL	
		Start	Finish					Sent	Rcvd		Sent	Rcvd
1												
	Notes											
2												
	Notes											
3												
	Notes											
4												
	Notes											
5												
	Notes											
6												
	Notes											
7												
	Notes											
8												
	Notes											
9												
	Notes											
10												
	Notes											
11												
	Notes											

Amateur Radio Station Log

Station Call sign:

QTH:

QRP - Mobile - Portable- Home

Antenna:

Date	Time UTC		Frequency (MHz)	Station	Mode	Power	Report		QTH	QSL		
	Start	Finish					Sent	Rcvd		Sent	Rcvd	
												1
Notes												
												2
Notes												
												3
Notes												
												4
Notes												
												5
Notes												
												6
Notes												
												7
Notes												
												8
Notes												
												9
Notes												
												10
Notes												
												11
Notes												

Amateur Radio Station Log

Station Call sign:

QRP - Mobile - Portable- Home

QTH:

Antenna:

	Date	Time UTC		Frequency (MHz)	Station	Mode	Power	Report		QTH	QSL	
		Start	Finish					Sent	Rcvd		Sent	Rcvd
1												
	Notes											
2												
	Notes											
3												
	Notes											
4												
	Notes											
5												
	Notes											
6												
	Notes											
7												
	Notes											
8												
	Notes											
9												
	Notes											
10												
	Notes											
11												
	Notes											

Amateur Radio Station Log

Station Call sign: | *QTH:*

QRP - Mobile - Portable- Home | *Antenna:*

Date	Time UTC		Frequency (MHz)	Station	Mode	Power	Report		QTH	QSL		
	Start	Finish					Sent	Rcvd		Sent	Rcvd	
												1
Notes												
												2
Notes												
												3
Notes												
												4
Notes												
												5
Notes												
												6
Notes												
												7
Notes												
												8
Notes												
												9
Notes												
												10
Notes												
												11
Notes												

Amateur Radio Station Log

Station Call sign:

QRP - Mobile - Portable- Home

QTH:

Antenna:

	Date	Time UTC		Frequency (MHz)	Station	Mode	Power	Report		QTH	QSL	
		Start	Finish					Sent	Rcvd		Sent	Rcvd
1												
	Notes											
2												
	Notes											
3												
	Notes											
4												
	Notes											
5												
	Notes											
6												
	Notes											
7												
	Notes											
8												
	Notes											
9												
	Notes											
10												
	Notes											
11												
	Notes											

Amateur Radio Station Log

Station Call sign:

QRP - Mobile - Portable- Home

QTH:

Antenna:

Date	Time UTC		Frequency (MHz)	Station	Mode	Power	Report		QTH	QSL		
	Start	Finish					Sent	Rcvd		Sent	Rcvd	
												1
Notes												
												2
Notes												
												3
Notes												
												4
Notes												
												5
Notes												
												6
Notes												
												7
Notes												
												8
Notes												
												9
Notes												
												10
Notes												
												11
Notes												

Amateur Radio Station Log

Station Call sign:

QTH:

QRP - Mobile - Portable- Home

Antenna:

	Date	Time UTC		Frequency (MHz)	Station	Mode	Power	Report		QTH	QSL	
		Start	Finish					Sent	Rcvd		Sent	Rcvd
1												
	Notes											
2												
	Notes											
3												
	Notes											
4												
	Notes											
5												
	Notes											
6												
	Notes											
7												
	Notes											
8												
	Notes											
9												
	Notes											
10												
	Notes											
11												
	Notes											

Amateur Radio Station Log

Station Call sign: *QTH:*

QRP - Mobile - Portable- Home *Antenna:*

Date	Time UTC		Frequency (MHz)	Station	Mode	Power	Report		QTH	QSL		
	Start	Finish					Sent	Rcvd		Sent	Rcvd	
												1
Notes												
												2
Notes												
												3
Notes												
												4
Notes												
												5
Notes												
												6
Notes												
												7
Notes												
												8
Notes												
												9
Notes												
												10
Notes												
												11
Notes												

Amateur Radio Station Log

Station Call sign: *QTH:*

QRP - Mobile - Portable- Home *Antenna:*

	Date	Time UTC		Frequency (MHz)	Station	Mode	Power	Report		QTH	QSL	
		Start	Finish					Sent	Rcvd		Sent	Rcvd
1												
	Notes											
2												
	Notes											
3												
	Notes											
4												
	Notes											
5												
	Notes											
6												
	Notes											
7												
	Notes											
8												
	Notes											
9												
	Notes											
10												
	Notes											
11												
	Notes											

Amateur Radio Station Log

Station Call sign:

QRP - Mobile - Portable- Home

QTH:

Antenna:

Date	Time UTC		Frequency (MHz)	Station	Mode	Power	Report		QTH	QSL		
	Start	Finish					Sent	Rcvd		Sent	Rcvd	
												1
Notes												
												2
Notes												
												3
Notes												
												4
Notes												
												5
Notes												
												6
Notes												
												7
Notes												
												8
Notes												
												9
Notes												
												10
Notes												
												11
Notes												

Amateur Radio Station Log

Station Call sign:

QTH:

QRP - Mobile - Portable- Home

Antenna:

	Date	Time UTC		Frequency (MHz)	Station	Mode	Power	Report		QTH	QSL	
		Start	Finish					Sent	Rcvd		Sent	Rcvd
1												
	Notes											
2												
	Notes											
3												
	Notes											
4												
	Notes											
5												
	Notes											
6												
	Notes											
7												
	Notes											
8												
	Notes											
9												
	Notes											
10												
	Notes											
11												
	Notes											

Amateur Radio Station Log

Station Call sign: *QTH:*

QRP - Mobile - Portable- Home *Antenna:*

Date	Time UTC		Frequency (MHz)	Station	Mode	Power	Report		QTH	QSL		
	Start	Finish					Sent	Rcvd		Sent	Rcvd	
												1
Notes												
												2
Notes												
												3
Notes												
												4
Notes												
												5
Notes												
												6
Notes												
												7
Notes												
												8
Notes												
												9
Notes												
												10
Notes												
												11
Notes												

Amateur Radio Station Log

Station Call sign:

QRP - Mobile - Portable- Home

QTH:

Antenna:

	Date	Time UTC		Frequency (MHz)	Station	Mode	Power	Report		QTH	QSL	
		Start	Finish					Sent	Rcvd		Sent	Rcvd
1												
	Notes											
2												
	Notes											
3												
	Notes											
4												
	Notes											
5												
	Notes											
6												
	Notes											
7												
	Notes											
8												
	Notes											
9												
	Notes											
10												
	Notes											
11												
	Notes											

Amateur Radio Station Log

Station Call sign:

QTH:

QRP - Mobile - Portable- Home

Antenna:

Date	Time UTC		Frequency (MHz)	Station	Mode	Power	Report		QTH	QSL		
	Start	Finish					Sent	Rcvd		Sent	Rcvd	
												1
Notes												
												2
Notes												
												3
Notes												
												4
Notes												
												5
Notes												
												6
Notes												
												7
Notes												
												8
Notes												
												9
Notes												
												10
Notes												
												11
Notes												

Amateur Radio Station Log

Station Call sign:

QRP - Mobile - Portable- Home

QTH:

Antenna:

	Date	Time UTC		Frequency (MHz)	Station	Mode	Power	Report		QTH	QSL	
		Start	Finish					Sent	Rcvd		Sent	Rcvd
1												
	Notes											
2												
	Notes											
3												
	Notes											
4												
	Notes											
5												
	Notes											
6												
	Notes											
7												
	Notes											
8												
	Notes											
9												
	Notes											
10												
	Notes											
11												
	Notes											

Amateur Radio Station Log

Station Call sign: *QTH:*

QRP - Mobile - Portable- Home *Antenna:*

Date	Time UTC		Frequency (MHz)	Station	Mode	Power	Report		QTH	QSL		
	Start	Finish					Sent	Rcvd		Sent	Rcvd	
												1
Notes												
												2
Notes												
												3
Notes												
												4
Notes												
												5
Notes												
												6
Notes												
												7
Notes												
												8
Notes												
												9
Notes												
												10
Notes												
												11
Notes												

Amateur Radio Station Log

Station Call sign:

QRP - Mobile - Portable- Home

QTH:

Antenna:

	Date	Time UTC		Frequency (MHz)	Station	Mode	Power	Report		QTH	QSL	
		Start	Finish					Sent	Rcvd		Sent	Rcvd
1												
	Notes											
2												
	Notes											
3												
	Notes											
4												
	Notes											
5												
	Notes											
6												
	Notes											
7												
	Notes											
8												
	Notes											
9												
	Notes											
10												
	Notes											
11												
	Notes											

Amateur Radio Station Log

Station Call sign: *QTH:*

QRP - Mobile - Portable- Home *Antenna:*

Date	Time UTC		Frequency (MHz)	Station	Mode	Power	Report		QTH	QSL		
	Start	Finish					Sent	Rcvd		Sent	Rcvd	
												1
Notes												
												2
Notes												
												3
Notes												
												4
Notes												
												5
Notes												
												6
Notes												
												7
Notes												
												8
Notes												
												9
Notes												
												10
Notes												
												11
Notes												

Amateur Radio Station Log

Station Call sign:

QTH:

QRP - Mobile - Portable- Home

Antenna:

	Date	Time UTC		Frequency (MHz)	Station	Mode	Power	Report		QTH	QSL	
		Start	Finish					Sent	Rcvd		Sent	Rcvd
1												
	Notes											
2												
	Notes											
3												
	Notes											
4												
	Notes											
5												
	Notes											
6												
	Notes											
7												
	Notes											
8												
	Notes											
9												
	Notes											
10												
	Notes											
11												
	Notes											

Amateur Radio Station Log

Station Call sign:

QTH:

QRP - Mobile - Portable- Home

Antenna:

Date	Time UTC		Frequency (MHz)	Station	Mode	Power	Report		QTH	QSL		
	Start	Finish					Sent	Rcvd		Sent	Rcvd	
												1
Notes												
												2
Notes												
												3
Notes												
												4
Notes												
												5
Notes												
												6
Notes												
												7
Notes												
												8
Notes												
												9
Notes												
												10
Notes												
												11
Notes												

Amateur Radio Station Log

Station Call sign: QTH:

QRP - Mobile - Portable- Home Antenna:

	Date	Time UTC		Frequency (MHz)	Station	Mode	Power	Report		QTH	QSL	
		Start	Finish					Sent	Rcvd		Sent	Rcvd
1												
	Notes											
2												
	Notes											
3												
	Notes											
4												
	Notes											
5												
	Notes											
6												
	Notes											
7												
	Notes											
8												
	Notes											
9												
	Notes											
10												
	Notes											
11												
	Notes											

Amateur Radio Station Log

Station Call sign:

QTH:

QRP - Mobile - Portable- Home

Antenna:

Date	Time UTC		Frequency (MHz)	Station	Mode	Power	Report		QTH	QSL		
	Start	Finish					Sent	Rcvd		Sent	Rcvd	
												1
Notes												
												2
Notes												
												3
Notes												
												4
Notes												
												5
Notes												
												6
Notes												
												7
Notes												
												8
Notes												
												9
Notes												
												10
Notes												
												11
Notes												

Amateur Radio Station Log

Station Call sign:

QTH:

QRP - Mobile - Portable- Home

Antenna:

	Date	Time UTC		Frequency (MHz)	Station	Mode	Power	Report		QTH	QSL	
		Start	Finish					Sent	Rcvd		Sent	Rcvd
1												
	Notes											
2												
	Notes											
3												
	Notes											
4												
	Notes											
5												
	Notes											
6												
	Notes											
7												
	Notes											
8												
	Notes											
9												
	Notes											
10												
	Notes											
11												
	Notes											

Amateur Radio Station Log

Station Call sign:

QTH:

QRP - Mobile - Portable- Home

Antenna:

Date	Time UTC		Frequency (MHz)	Station	Mode	Power	Report		QTH	QSL		
	Start	Finish					Sent	Rcvd		Sent	Rcvd	
												1
Notes												
												2
Notes												
												3
Notes												
												4
Notes												
												5
Notes												
												6
Notes												
												7
Notes												
												8
Notes												
												9
Notes												
												10
Notes												
												11
Notes												

Amateur Radio Station Log

Station Call sign:

QRP - Mobile - Portable- Home

QTH:

Antenna:

	Date	Time UTC		Frequency (MHz)	Station	Mode	Power	Report		QTH	QSL	
		Start	Finish					Sent	Rcvd		Sent	Rcvd
1												
	Notes											
2												
	Notes											
3												
	Notes											
4												
	Notes											
5												
	Notes											
6												
	Notes											
7												
	Notes											
8												
	Notes											
9												
	Notes											
10												
	Notes											
11												
	Notes											

Amateur Radio Station Log

Station Call sign:

QTH:

QRP - Mobile - Portable- Home

Antenna:

Date	Time UTC		Frequency (MHz)	Station	Mode	Power	Report		QTH	QSL	
	Start	Finish					Sent	Rcvd		Sent	Rcvd
												1
Notes											
												2
Notes											
												3
Notes											
												4
Notes											
												5
Notes											
												6
Notes											
												7
Notes											
												8
Notes											
												9
Notes											
												10
Notes											
												11
Notes											

Amateur Radio Station Log

Station Call sign:

QTH:

QRP - Mobile - Portable- Home

Antenna:

Date	Time UTC		Frequency (MHz)	Station	Mode	Power	Report		QTH	QSL	
	Start	Finish					Sent	Rcvd		Sent	Rcvd
1											
Notes											
2											
Notes											
3											
Notes											
4											
Notes											
5											
Notes											
6											
Notes											
7											
Notes											
8											
Notes											
9											
Notes											
10											
Notes											
11											
Notes											

Amateur Radio Station Log

Station Call sign:

QRP - Mobile - Portable- Home

QTH:

Antenna:

Date	Time UTC		Frequency (MHz)	Station	Mode	Power	Report		QTH	QSL		
	Start	Finish					Sent	Rcvd		Sent	Rcvd	
												1
Notes												
												2
Notes												
												3
Notes												
												4
Notes												
												5
Notes												
												6
Notes												
												7
Notes												
												8
Notes												
												9
Notes												
												10
Notes												
												11
Notes												

Amateur Radio Station Log

Station Call sign:

QRP - Mobile - Portable- Home

QTH:

Antenna:

	Date	Time UTC		Frequency (MHz)	Station	Mode	Power	Report		QTH	QSL	
		Start	Finish					Sent	Rcvd		Sent	Rcvd
1												
	Notes											
2												
	Notes											
3												
	Notes											
4												
	Notes											
5												
	Notes											
6												
	Notes											
7												
	Notes											
8												
	Notes											
9												
	Notes											
10												
	Notes											
11												
	Notes											

Amateur Radio Station Log

Station Call sign: *QTH:*

QRP - Mobile - Portable- Home *Antenna:*

Date	Time UTC		Frequency (MHz)	Station	Mode	Power	Report		QTH	QSL		
	Start	Finish					Sent	Rcvd		Sent	Rcvd	
												1
Notes												
												2
Notes												
												3
Notes												
												4
Notes												
												5
Notes												
												6
Notes												
												7
Notes												
												8
Notes												
												9
Notes												
												10
Notes												
												11
Notes												

Amateur Radio Station Log

Station Call sign:

QTH:

QRP - Mobile - Portable- Home

Antenna:

	Date	Time UTC		Frequency (MHz)	Station	Mode	Power	Report		QTH	QSL	
		Start	Finish					Sent	Rcvd		Sent	Rcvd
1												
	Notes											
2												
	Notes											
3												
	Notes											
4												
	Notes											
5												
	Notes											
6												
	Notes											
7												
	Notes											
8												
	Notes											
9												
	Notes											
10												
	Notes											
11												
	Notes											

Amateur Radio Station Log

Station Call sign: *QTH:*

QRP - Mobile - Portable- Home *Antenna:*

Date	Time UTC		Frequency (MHz)	Station	Mode	Power	Report		QTH	QSL		
	Start	Finish					Sent	Rcvd		Sent	Rcvd	
												1
Notes												
												2
Notes												
												3
Notes												
												4
Notes												
												5
Notes												
												6
Notes												
												7
Notes												
												8
Notes												
												9
Notes												
												10
Notes												
												11
Notes												

Amateur Radio Station Log

Station Call sign:

QTH:

QRP - Mobile - Portable- Home

Antenna:

	Date	Time UTC		Frequency (MHz)	Station	Mode	Power	Report		QTH	QSL	
		Start	Finish					Sent	Rcvd		Sent	Rcvd
1												
	Notes											
2												
	Notes											
3												
	Notes											
4												
	Notes											
5												
	Notes											
6												
	Notes											
7												
	Notes											
8												
	Notes											
9												
	Notes											
10												
	Notes											
11												
	Notes											

Amateur Radio Q Codes

Code	Question	Answer or Statement
QNI	May I join the net?	You may check in...
QRA	What is the call sign of your station?	The call sign of my station is ...
QRG	Will you tell me my exact frequency?	Your exact frequency is ... kHz (or MHz).
QRH	Does my frequency vary?	Your frequency varies.
QRI	How is the tone of my transmission?	"The tone of your transmission is (1. Good; 2. Variable; 3. Bad)"
QRJ	How many voice contacts do you want to make?	I want to make ... voice contacts.
QRK	What is the readability of my signals?	"The readability of your signals is ... (1 to 5)."
QRL	Are you busy?	I am busy. Please do not interfere.
QRM	Do you have interference? [from other stations]	I have interference.
QRN	Are you troubled by static?	I am troubled by static.
QRO	Shall I increase power?	Increase power.
QRP	Shall I decrease power?	Decrease power.
QRQ	Shall I send faster?	Send faster (... wpm)
QRS	Shall I send more slowly?	Send more slowly (... wpm)
QRT	Shall I cease or suspend operation?	I am suspending operation.
QRU	Have you anything for me?	I have nothing for you.
QRV	Are you ready?	I am ready.
QRW	Shall I inform ... that you are calling him on ... kHz (or MHz)?	Please inform ... that I am calling him on ... kHz (or MHz).
QRX	Shall I standby / When will you call me again?	Please standby / I will call you again at ... (hours) on ... kHz (or MHz)
QRZ	Who is calling me?	You are being called by ... on ... kHz (or MHz)
QSA	What is the strength of my signals (or those of ...)?	The strength of your signals (or those of ...) is ... (1 to 5).
QSB	Are my signals fading?	Your signals are fading.
QSD	Is my keying defective?	Your keying is defective.
QSG	Shall I send ... messages at a time?	Send ... messages at a time.

Code	Question	Answer or Statement
QSK	Can you hear me between your signals?	I can hear you between my signals.
QSL	Can you acknowledge receipt?	I am acknowledging receipt.
QSM	Shall I repeat the last message which I sent you?	Repeat the last message which you sent me.
QSN	Did you hear me on .. kHz (or MHz)?	I did hear you on ... kHz (or MHz).
QSO	Can you communicate with ... direct or by relay?	I can communicate with ... direct (or by relay through ...).
QSP	Will you relay a message to ...?	I will relay a message to ... .
QSR	Do you want me to repeat my call?	Please repeat your call; I did not hear you.
QSS	What working frequency will you use?	I will use the working frequency ... kHz (or MHz).
QST	Should I repeat the prior message to all amateurs I contact?	Here follows a broadcast message to all amateurs.
QSU	Shall I send or reply on this frequency (or on ... kHz (or MHz))?	Send or reply on this frequency (or on ... kHz (or MHz)).
QSW	Will you send on this frequency (or on ... kHz (or MHz))?	I am going to send on this frequency (or on ... kHz (or MHz)).
QSX	Will you listen to ... (call sign(s) on ... kHz (or MHz))?	I am listening to ... (call sign(s) on ... kHz (or MHz))
QSY	Shall I change to transmission on another frequency?	Change to transmission on another frequency (or on ... kHz (or MHz)).
QSZ	Shall I send each word or group more than once?	Send each word or group twice (or ... times).
QTC	How many messages have you to send?	I have ... messages for you.
QTH	"What is your position in latitude and longitude?"	My position is ... Latitude ... longitude
QTR	What is the correct time?	The correct time is ... hours
QTU	At what times are you operating?	I am operating from ... to ... hours.
QTX	Will you keep your station open for further communication with me until further notice?	I will keep my station open for further communication with you until further notice (or until ... hours).
QUA	Have you news of ... (call sign)?	Here is news of ... (call sign).
QUC	What is the number (or other indication) of the last message you received from me?	The number (or other indication) of the last message I received from you is ...
QUD	Have you received the urgency signal sent by ... (call sign)?	I have received the urgency signal sent by ... (call sign) at ... hours.
QUE	Can you speak in ... (language); if so, on what frequencies?	I can speak in ... (language) on ... kHz (or MHz).
QUF	Have you received the distress signal sent by ... (call sign)?	I have received the distress signal sent by ... (call sign) at ... hours.

RST System

No.	Meaning		
	R	S	T
9		Extremely strong signals	"Perfect tone, no trace of ripple or modulation of any kind"
8		Strong signals	"Near perfect tone, slight trace of modulation"
7		Moderately strong signals	"Near pure tone, trace of ripple modulation"
6		Good signals	"Filtered tone, definite trace of ripple modulation"
5	Perfectly readable	Fairly good signals	"Filtered rectified a.c. but strongly ripple-modulated"
4	Readable with practically no difficulty	Fair signals	"Rough note, some trace of filtering"
3	Readable with considerable difficulty	Weak signals	"Rough a.c. tone, rectified but not filtered"
2	"Barely readable, occasional word distinguishable"	Very weak signals	"Very rough a.c., very harsh and broad"
1	Unreadable	Faint, signals barely perceptible	"Sixty cycle a.c or less, very rough and broad"

Radiotelephony and Morse code

Character	Morse Code	Telephony	"International Pronunciation"
A	·–	Alfa	(AL-FAH)
B	–···	Bravo	(BRAH-VOH)
C	–·–·	Charlie	(CHAR-LEE)
D	–··	Delta	(DELL-TAH)
E	·	Echo	(ECK-OH)
F	··–·	Foxtrot	(FOKS-TROT)
G	––·	Golf	(GOLF)
H	····	Hotel	(HOH-TEL)
I	··	India	(IN-DEE-AH)
J	·–––	Juliett	(JEW-LEE-ETT)
K	–·–	Kilo	(KEY-LOH)
L	·–··	Lima	(LEE-MAH)
M	––	Mike	(MIKE)

Character	Morse Code	Telephony	"International Pronunciation"
N	–·	November	(NO-VEM-BER)
O	–––	Oscar	(OSS-CAH)
P	·––·	Papa	(PAH-PAH)
Q	––·–	Quebec	(KEH-BECK)
R	·–·	Romeo	(ROW-ME-OH)
S	···	Sierra	(SEE-AIR-RAH)
T	–	Tango	(TANG-GO)
U	··–	Uniform	(YOU-NEE-FORM)
V	···–	Victor	(VIK-TAH)
W	·––	Whiskey	(WISS-KEY)
X	–··–	Xray	(ECKS-RAY)
Y	–·––	Yankee	(YANG-KEY)
Z	––··	Zulu	(ZOO-LOO)

1	·––––	One	(WUN)
2	··–––	Two	(TOO)
3	···––	Three	(TREE)
4	····–	Four	(FOW-ER)
5	·····	Five	(FIFE)

6	–····	Six	(SIX)
7	––···	Seven	(SEV-EN)
8	–––··	Eight	(AIT)
9	––––·	Nine	(NIN-ER)
0	–––––	Zero	(ZEE-RO)

International Call Sign Series

Call Sign Prefix	Country	Worked on	CQ Zone	ITU Zone
A2	Botswana	❑	38	57
A3	Tonga Island	❑	32	62
A4	Oman	❑	21	39
A5	Bhutan	❑	22	41
A6	United Arab Emirates	❑	21	39
A7	Qatar	❑	21	39
A9	Bahrain	❑	21	39
AA → AK	USA	❑	3, 4, 5	6, 7, 8
AP → AS	Pakistan	❑	21	41
BS7	Scarborough Reef	❑	27	50
BV	Taiwan	❑	24	44
BV9P	Pratas	❑	24	44
BX → BZ	China	❑	23, 24	42, 43, 44
C2	Nauru	❑	31	65
C3	Andorra	❑	14	27
C5	Gambia	❑	35	46
C6	Bahamas	❑	8	11
C9	Mozambique	❑	37	53
CE	Chili	❑	12	14, 15, 16
CE9, KC4	Antarctica	❑	13	73
CEo	Easter Island	❑	12	63
CEo	San Felix Island	❑	12	14
CEo	Juan Fernandez Island	❑	12	14
CM, CO	Cuba	❑	8	11
CN	Morocco	❑	33	37
CP	Bolivia	❑	12	12, 13, 14
CT	Portugal	❑	14	37
CT3	Madeira Island	❑	33	36
CU	Azores Island	❑	14	36
CV → CX	Uruguay	❑	13	14
CYo	Sable Island	❑	5	9
CY9	St Paul Island	❑	5	9
D2, D3	Angola	❑	36	52
D4	Cape Verde	❑	35	46
D6	Comoros	❑	39	53
DA → DM	Germany	❑	14	28
DU, DZ	Philippines	❑	27	50
E3	Erythrea	❑	37	48
E4	Palestina	❑	20	39
EA → EH	Spain	❑	14	37
EA6 → EH6	Balearic Islands	❑	14	37
EA8 → EH8	Canary Island	❑	33	36
EA9 → EH9	Ceuta & Melilla	❑	33	37

Call Sign Prefix	Country	Worked on	CQ Zone	ITU Zone
EI, EJ	Ireland	❑	14	27
EK	Armenia	❑	21	29
EL	Liberia	❑	35	46
EM, EO	Ukraine	❑	16	29
EP, EQ	Iran	❑	21	40
ER	Moldova	❑	16	29
ES	Estonia	❑	15	29
ET	Ethiopia	❑	37	48
EU, EV, EW	Belarus	❑	16	29
EX	Kyrgystan	❑	17	30, 31
EY	Tajikistan	❑	17	30
EZ	Turkmenistan	❑	17	30
FA → FF	France	❑	14	27
FG	Guadeloupe	❑	8	11
FH	Mayotte	❑	39	53
FJ, FS	French St. Martin	❑	8	11
FK	New Caledonia	❑	32	56
FM	Martinique	❑	8	11
FO	French Polynesia	❑	32	63
FOo	Clipperton Isl.	❑	7	10
FOo	Marquesas	❑	31	63
FOo	Austral Islands	❑	32	63
FP	St. Pierre & Miquelon	❑	5	9
FR	Reunion	❑	39	53
FR/J	Juan de Nova	❑	39	53
FR/G	Glorioso	❑	39	53
FR/T	Tromelin	❑	39	53
FT8W	Crozet Island	❑	39	68
FT8X	Kerguelen Island	❑	39	68
FT8Z	Amsterdam & St. Paul Island	❑	39	68
FW	Wallis & Futuna Island	❑	32	62
FY	French Guiana	❑	9	12
G, GX, M	England	❑	14	27
GD, GT, MD	Isle of Man	❑	14	27
GI, GN, MI	Northern Ireland	❑	14	27
GJ, GH, MJ	Jersey	❑	14	27
GM, GS, MM	Scotland	❑	14	27
GU, GP, MU	Guernsey	❑	14	27
GW, GC, MW	Wales	❑	14	27
H4	Solomon Island	❑	28	51
H4o	Temotu Province	❑	32	51
HA, HG	Hungary	❑	14	28
HB	Switzerland	❑	14	28

International Call Sign Series

Call Sign Prefix	Country	Worked on	CQ Zone	ITU Zone
HBo	Liechtenstein	❑	14	28
HC, HD	Ecuador	❑	10	12
HC8, HD8	Galapagos Island	❑	10	12
HH	Haiti	❑	8	11
HI	Dominican Republic	❑	8	11
HK, HJ	Colombia	❑	9	12
HKoM	Malpelo Island	❑	9	12
HKoS	San Andres & Providencia	❑	7	11
HL, HM, DS	South Korea	❑	25	44
HP, HO, 3E	Panama	❑	7	11
HR, HQ	Honduras	❑	7	11
HS, E2	Thailand	❑	26	49
HV	Vatican	❑	15	28
HZ, 7Z	Saudi Arabia	❑	21	39
I	Italy	❑	15,33	28, 37
ISo, IMo	Sardinia	❑	15	28
J2	Djibouti	❑	37	48
J3	Grenada	❑	8	11
J5	Guinea Bissau	❑	35	46
J6	St. Lucia	❑	8	11
J7	Dominica	❑	8	11
J8	St. Vincent	❑	8	11
JA → JS	Japan	❑	25	45
JD1M	Minami Torishima	❑	27	90
JD1O	Ogasawara	❑	27	45
JT, JV	Mongolia	❑	23	32
JW	Svalbard Island	❑	40	18
JX	Jan Mayen Island	❑	40	18
JY	Jordan	❑	20	39
K	USA	❑	3, 4, 5	6, 7, 8
T8	Palau	❑	27	64
KG4	Guantanamo Bay	❑	8	11
KH1	Baker & Howland Island	❑	31	61
KH2	Guam	❑	27	64
KH3	Johnston Island	❑	31	61
KH4	Midway Island	❑	31	61
KH5	Palmyra & Jarvis Island	❑	31	61, 62
KH5K	Kingman Reef	❑	31	61
KH6, KH7	Hawaii Island	❑	31	61
KH7K	Kure Island	❑	31	61
KH8	American Samoa	❑	32	62
KH9	Wake Island	❑	31	65
KH0	Mariana Island	❑	27	64

Call Sign Prefix	Country	Worked on	CQ Zone	ITU Zone
KL7	Alaska	❑	1	01, 02
KP1	Navassa Island	❑	8	11
KP2	Virgin Island	❑	8	11
KP3, KP4	Puerto Rico	❑	8	11
KP5	Desecheo Island	❑	8	11
LA, LN	Norway	❑	14	18
L, LU, LW	Argentina	❑	13	14, 15, 16
LX	Luxembourg	❑	14	27
LY	Lithuania	❑	15	29
LZ	Bulgaria	❑	20	28
N	USA	❑	3, 4, 5	6, 7, 8
OA, OC	Peru	❑	10	12
OD	Lebanon	❑	20	39
OE	Austria	❑	15	28
OG, OH, OI	Finland	❑	15	18
OHo	Aland Island	❑	15	18
OJo, OHoM	Market Reef	❑	15	18
OK, OL	Czech Republic	❑	15	28
OM	Slovak Republic	❑	15	28
ON → OT	Belgium	❑	14	27
OX	Greenland	❑	40	05, 75
OY	Faroe Island	❑	14	18
OZ	Denmark	❑	14	28
P2	Papua New Guinea	❑	28	51
P4	Aruba	❑	9	11
P5	North Korea	❑	25	44
PA → PI	The Netherlands	❑	14	27
PJ2	Curacao	❑	9	11
PJ4	Bonaire	❑	9	11
PJ5, PJ6	Saba and St. Eustatius	❑	8	11
PJ7	Sint Marteen	❑	8	11
PP → PY	Brazil	❑	11	12 → 15
PPo, PYoF	Fernando de Noronha	❑	11	13
PPo, PYoS	St. Peter & St Paul's Rocks	❑	11	13
PPo PYoT	Trindade & Martim Vaz. Isl.	❑	11	15
PZ	Surinam	❑	9	12
RA → RZ	Russia	❑	16 → 19	19, 21, 27, 29, 30, 33
RW2	Kaliningrad	❑	15	29
R1FJ	Franz Josef Land	❑	40	75
R1MV	Malyj Vysotskij Island	❑	16	29
S2	Bangladesh	❑	22	41
S5	Slovenia	❑	15	28
S7	Seychelles	❑	39	53

International Call Sign Series

Call Sign Prefix	Country	Worked on	CQ Zone	ITU Zone
S9	Sao Tome & Principe	❏	36	47
S0	Western Sahara	❏	33	46
SA → SM	Sweden	❏	14	18
SN → SR	Poland	❏	15	28
ST	Sudan	❏	34	48
SU	Egypt	❏	34	38
SV, SZ	Greece	❏	20	28
SV5	Dodecanese	❏	20	28
SV9	Crete	❏	20	28
SV2A	Mount Athos	❏	20	28
T2	Tuvalu	❏	31	65
T30	West Kiribati	❏	31	65
T31	Cent. Kiribati	❏	31	62
T32	East Kiribati	❏	31	61 → 63
T33	Banaba Island	❏	31	65
T5	Somalia	❏	37	48
T7	San Marino	❏	15	28
T9	Bosnia - Herzegovina	❏	15	28
TA, TB, TC	Turkey	❏	20	39
TF	Iceland	❏	40	17
TG, TD	Guatemala	❏	7	11
TI, TE	Costa Rica	❏	7	11
TI9	Cocos Island	❏	7	11
TJ	Cameroon	❏	36	47
TK	Corsica	❏	15	28
TL	Centr. African Rep.	❏	36	47
TM	France	❏	14	27
TN	Congo	❏	36	52
TR	Gabon	❏	36	52
TT	Chad	❏	36	47
TU	Ivory Coast	❏	35	46
TY	Benin	❏	35	46
TX	Chesterfield Islands	❏	39	53
TZ	Mali	❏	35	46
UA → UI 1 → 7	European Russia	❏	16	19, 29, 30
UA2, RW2	Kaliningrad	❏	15	29
UA → UI 8 → 0	Asiatic Russia	❏	16 → 19	21, 27, 30, 33
UJ → UM	Uzbekistan	❏	17	30
UN → UQ	Kazakhstan	❏	17	29, 30, 31
UR → UZ	Ukraine	❏	16	29
V2	Antigua & Barbuda	❏	8	11
V3	Belize	❏	7	11
V4	St. Kits, Nevis	❏	8	11

Call Sign Prefix	Country	Worked on	CQ Zone	ITU Zone
V5	Namibia	❏	38	57
V6	Micronesia	❏	27	65
V7	Marshall Island	❏	31	65
V8	Brunei	❏	28	54
VA → VG, VO, VY	Canada	❏	01 → 05	02 → 09
VK	Australia	❏	29, 30	55, 58, 59
VK9C	Cocos Keeling Island	❏	29	54
VK9L	Lord Howe Island	❏	30	60
VK9M	Mellish Reef	❏	30	56
VK9N	Norfolk Island	❏	32	60
VK9W	Willis Island	❏	30	55
VK9X	Christmas Island	❏	29	54
VK0	Heard Island	❏	39	68
VK0	Macquarie Island	❏	30	60
VP2E	Anguilla	❏	8	11
VP2M	Montserrat	❏	8	11
VP2V	British Virgin Island	❏	8	11
VP5	Turks & Caicos Island	❏	8	11
VP6	Pitcairn Island	❏	32	63
VP8	Falkland Island	❏	13	16
VP8, LU	South Georgia Island	❏	13	73
VP8, LU	South Orkney Island	❏	13	73
VP8, LU	South Sandwich Island	❏	13	73
VP8, LU, HF0, CE9	South Shetland Island	❏	13	73
VP9	Bermuda	❏	5	11
VQ9	Chagos	❏	39	41
VS6, VR2	Hong Kong	❏	24	44
VU	India	❏	22	41
VU4	Andaman & Nicobar Island	❏	26	49
VU7	Laccadive Island	❏	22	41
XA → XI	Mexico	❏	6	10
XA4 → XI4	Revillagigedo Islands	❏	6	10
W	USA	❏	3, 4, 5	6, 7, 8
XT	Burkina Faso	❏	35	46
XU	Cambodia	❏	26	49
XV	Vietnam	❏	26	49
XW	Laos	❏	26	49
XX9	Macao	❏	24	44
XZ, XY	Myanmar	❏	26	49
YA	Afghanistan	❏	21	40
YB → YH	Indonesia	❏	28	51 → 54
YI	Iraq	❏	21	39
YJ	Vanuatu	❏	32	56

International Call Sign Series

Call Sign Prefix	Country	Worked on	CQ Zone	ITU Zone
YK	Syria	❑	20	39
YL	Latvia	❑	15	29
YN	Nicaragua	❑	7	11
YO → YR	Romania	❑	20	28
YS	El Salvador	❑	7	11
YU, YT	Serbia	❑	15	28
YV → YY	Venezuela	❑	9	12
YVo	Aves Island	❑	8	11
Z2	Zimbabwe	❑	38	53
Z3	Macedonia	❑	15	28
ZA	Albania	❑	15	28
ZB2	Gibraltar	❑	14	37
ZC4	British Base on Cyprus	❑	20	39
ZD7	St. Helena	❑	36	66
ZD8	Ascension Island	❑	36	66
ZD9	Tristan Da Cunha Island	❑	38	66
ZF	Cayman Island	❑	8	11
ZK1N	Norht Cook Island	❑	32	62
ZK1S	South Cook Island	❑	32	62
ZK2	Niue	❑	32	62
ZK3	Tokelau	❑	31	62
ZL, ZM	New Zealand	❑	32	60
ZL7	Chatham Island	❑	32	60
ZL8	Kermadec Island	❑	32	60
ZL9	Auckland & Campbell Island	❑	32	60
ZP	Paraguay	❑	11	14
ZR → ZU	South Africa	❑	38	57
ZS8	Marion & Pr. Edward Island	❑	38	57
1A	Military Order of Malta	❑	15	28
1S, 9Mo	Spratly Island	❑	26	50
3A	Monaco	❑	14	27
3B6, 3B7	Agalega & St. Brandon	❑	39	53
3B8	Mauritius	❑	39	53
3B9	Rodrigues Island	❑	39	53
3C	Equatorial Guinea	❑	36	47
3Co	Annobon Island	❑	36	52
3D2	Fiji Islands	❑	32	56
3D2C	Conway Reef	❑	32	56
3D2R	Rotuma Island	❑	32	56
3DAo, 3D6	Swaziland	❑	38	57
3V	Tunisia	❑	33	37
3W	Vietnam	❑	26	49
3X	Guinea	❑	35	46

Call Sign Prefix	Country	Worked on	CQ Zone	ITU Zone
3Y5	Bouvet	❑	38	67
3Yo	Peter 1 Island	❑	12	72
4K, 4J	Azerbaijan	❑	21	29
4L	Georgia	❑	21	29
4P → 4S	Sri Lanka	❑	22	41
4U1ITU	I.T.U. Geneva	❑	14	28
4U1UN	United Nations HQ	❑	5	8
4W	Timor Leste	❑	28	54
4X, 4Z	Israel	❑	20	39
5A	Libya	❑	34	38
5B, C4, P3	Cyprus	❑	20	39
5H, 5I	Tanzania	❑	37	53
5N, 5O	Nigeria	❑	35	46
5R, 5S	Madagascar	❑	39	53
5T	Mauritania	❑	35	46
5U	Niger	❑	35	46
5V	Togo	❑	35	46
5W	Western Samoa	❑	32	62
5X	Uganda	❑	37	48
5Y, 5Z	Kenya	❑	37	48
6V, 6W	Senegal	❑	35	46
6Y	Jamaica	❑	8	11
7O	Yemen	❑	21	39
7P	Lesotho	❑	38	57
7Q	Malawi	❑	37	53
7T → 7Y	Algeria	❑	33	37
8P	Barbados	❑	8	11
8Q	Maldives	❑	22	41
8R	Guyana	❑	9	12
9A	Croatia	❑	15	28
9G	Ghana	❑	35	46
9H	Malta	❑	15	28
9J, 9I	Zambia	❑	36	53
9K	Kuwait	❑	21	39
9L	Sierra Leone	❑	35	46
9M2, 4	West Malaysia	❑	28	54
9M6, 8	East Malaysia	❑	28	54
9N	Nepal	❑	22	42
9Q, 9T	Dem. Rep. of Congo	❑	36	52
9U	Burundi	❑	36	52
9V	Singapore	❑	28	54
9X	Rwanda	❑	36	52
9Y, 9Z	Trinidad & Tobago	❑	9	11

Notes

Printed by Amazon Italia Logistica S.r.l.
Torrazza Piemonte (TO), Italy

59575203R00075